Astronomy

Discover The Amazing Truth About New Galaxies, Worm Holes, Black Holes And The Latest Discoveries In Astronomy

Cover image courtesy of Gianni – Flickr - https://www.flickr.com/photos/xamad/518876976/

Table of Contents

Do You Want More Books?

How would you like books arriving in your inbox each week?

Don't worry they are FREE!

We publish books on all sorts of non-fiction niches and send them out to our subscribers each week to spread the love.

All you have to do is sign up and you're good to go!

Just go to the link at the end of this book, sign up, sit back and wait for your book downloads to arrive!

We couldn't have made it any easier! Enjoy!

Introduction

I want to thank you and congratulate you for purchasing the book, "Astronomy: *Discover The Amazing Truth About New Galaxies Worm Holes Black Holes And The Latest Discoveries In Astronomy*"

Have you ever looked into the heavens above and into the night and wondered about the curious little fireflies in the sky called stars? Ever wondered how they got there? How big are they? How far away are they? What is there in the centre of our galaxy?

These are the same questions asked by people as ancient as the Egyptians and the Babylonians. These were followed by their successors in Europe and now these initial few musings have developed into much deeper questions such as what is the nature of Dark Matter? Where and how did life begin if not on earth? What is the ultimate fate of the universe? What is the nature of gravity?

All of these challenging and exciting questions are addressed under a universal nomenclature called astronomy. Astronomy has been considered an ancient interdisciplinary study combining physics, chemistry and biology to study the space outside of earth. Astronomy is a powerful tool that shows our significance as a life sustaining planet and our insignificance as a dew drop in a vast ocean of stars.

We seek to give the 101 on an enormous field of study that is pursued by millions of amateurs at home and scientists in laboratories around the world. You will be learning about the entities that inhabit the dark sky.

You will learn how to identify them and what it means to be an Astronomer. Astronomy has enormous theoretical value and there will be quite a bit of interesting theory.

The book will introduce to you the basics of astronomy by exploring it starting from the nearest of the cosmological wonders such as the Sun, the Moon and the other entities of the night sky such as the various planets of the Solar System and the myriad constellations.

We will then take a little peak into the more modern and exciting discoveries such as worm holes, black holes, dark matter and other habitants of the universe.

And towards the end you can discover how to further your knowledge and contribute to this field, as some of the greatest astronomical findings have been made by amateurs.

So let us start by exploring how all of this came about.

Thanks again for purchasing this book, I hope you enjoy it!

Chapter 1: A Brief History Of Astronomy

Astronomy probably has one of the most ancient and monumental histories of all sciences. We have always wanted to know more about the black beyond the night and the bright sky of the day.

Astronomy started out as a primary field of observation for insight into the weather and the heavens. Usually priests and other religious figures were entrusted in reading signs to predict the seasons and the moods of Gods. And as we got more involved our study also became more organized and we soon began to build tools to be able understand and measure the sun, moon and the other stars more clearly.

Some of the earliest professional astronomers in this regard were the ancient Mesopotamians, Indians, Chinese and almost the rest of the world showed so much interest and understanding of space such that even now we are just confirming some of the findings.

Some of the instruments they built were the astrolabe, the quadrant, the armillary sphere, the great mural quadrant and other similar instruments to observe and record important information such as the angle of the heavenly bodies with respect to us, the various relative positions across their respective calendars and celestial co-ordinates.

The ancients even built extraordinary observatories all around the world and some of these are truly inspiring and stand testament to their depth of understanding and passion. Some of these are the Jantar Mantar observatory in Jaipur, India and the "El Caracol" observatory temple at Chichen Itza, Mexico.

Astronomy is a continuously evolving field and some of the ground breaking discoveries they made were calculating the radius of the Earth, the size and distance of the Sun and the Moon and some even mapped out the surface of Mars.

After this era, the next major developments in astronomy were made in Europe. During the early 17th century a revolutionary invention occurred that changed the limits of our horizon.

The invention of the telescope launched and continues to

launch some of the most inspiring minds of the world such as Nicolaus Copernicus, Galileo Gelilei and Johannes Kepler. They came up with the ideas such as heliocentric model, discovered new planets and made more accurate predictions of the motions of the Moon and the Sun.

After them, astronomy took off in leaps and bounds. The shift in interests was more intense as people began to unravel the different spectrums and observed and looked into the sky with even greater intensity. Isaac Newton, Einstein and Stephen Hawking are just some of the greats that helped us discover Gravity, Theory of Relativity and Singularity.

It would be ridiculous to ignore the close relationship between astronomy and physics as one studies the history of this great science. Physics is practically the theoretical canvas on which astronomy is made.

They have a symbiotic relationship in the sense that observations in astronomy help and prove theoretical predictions in physics and the theoretical explanations of physics have explained astronomical events.

Confirmation of the theory of relativity through the observation of the eclipse being a famous explanation of the former relation and the reason behind the blinking of stars is a suitable example of the latter relation.

To talk about the history of astronomy is in itself a separate field of study and that itself could take several volumes of books. One cannot do justice to this field by ignoring its history because to understand astronomy you have to understand and live its evolution. Remember that you would be exploring much the same way the ancients have and it pays dividends to learn about them.

Hence we shall start off with an exploration of our source of life on Earth, the Sun.

Chapter 2: The Sun

The sun has been the source of life on earth and we simply would not exist without it. Observing the Sun might be a little tricky as looking at it directly is very harmful and hard. But ancients have even built clocks that operate on the shadow made due to the light from the sun and have made the earliest of calendars based on its cycle through the day sky.

The sun is at the centre of our solar system and it was called "Helios" by the ancient Greeks. The sun is about 4.5 billion years old and it is what we call a second or even third generation star.

Basically in terms of star life, our sun is in the middle of its projected lifespan. It will eventually expand as its fuel of hydrogen runs out and turn into a massive red giant. At this state it will be several times bigger than it is now.

Finally, it will die as a black or brown dwarf as it completely exhausts everything. This brief explanation is the life sequence of a Main Sequence star. There are variations in the scale and duration of the above mentioned events in the sequence, some of which we will explore further.

It is approximately at the inner rim of the Orion arm of our Milky Way galaxy. The sun also orbits the centre of our galaxy at a distance of about 25000 light years. A Light Year is the distance travelled by light in a vacuum for a period of one year which turns out to be almost 9.5 trillion kilometres.

It is mainly composed of hydrogen at its core but it also burns heavier elements such as helium and heavier metals. The way it generates an enormous amount of energy is through internal fusion. Our sun, like all other stars, was born from out of the gaseous remnants of a supernova which the violent death of a star releases as it burns out its fuel of hydrogen.

It is actually white in color and it follows a path across the horizon which is around 20 days. It was used a basic guide in organizing the months.

One curious feature of the sun is solar spots. These are generally localized spots on the surface where the temperature is much lower than it should be. These regions have massive magnetic energy fluxes and are suspected to be the reason behind solar flares and coronal mass ejections.

This is of particular importance due to the destructive nature of the solar winds that are produced. These solar winds cause what are called solar storms; these storms have in the past played havoc with communication and the transmission of electricity.

These sunspots roughly co relate to the 11 year solar cycle, the time the sun takes to rotate about its own axis. Essentially the next time a sunspot is seen we are in trouble.

The contribution of the sun to studying astronomy is immense. It is used as a common reference point. In fact the distance from the centre of the earth to the centre of the sun is classified as one astronomical unit or AU is often more preferably used to express distances than light year.

The sun, despite being so close and observable is primarily studied through the use of spectroscopy. It also holds many important mysteries such as how is the temperature of the outer layers of the atmosphere of the sun is much higher than the temperature of the surface of the sun.

Scientists are still unable to figure out why this is so and the discovery of the mechanism that causes this anomaly could be the key to efficient energy transmission. This is known as Coronal Heating Problem and is an active area of research.

Such phenomenon were discovered by numerous space flights such as the Pioneers mission and continues to attract a lot of visitors such as the Aditya which is scheduled to visit the sun sometime around 2015-2016.

The sun is an inspirational light form, from its explosive core to its mysterious outer regions, on that note we shall get you introduced to our nearest cosmic neighbour, the Moon.

Other Unbelievable Facts About the Sun

- Unknown to many, our sun – and our whole solar system – is moving! In fact, the speed of the sun is much faster than the speed of the earth's rotation! If the earth spins at a speed of 1673 kilometres per hour, our sun moves at the speed of 13200 kilometres per hour! You might be asking: if the sun is moving, where will it go?

Nowhere in particular. The truth is, our solar system is revolving around the center of the Milky Way, but to be able to complete JUST ONE revolution, we have to wait 230 million years.

- While it is trivial to know the exact measurement of the sun, which is approximately 1.4 million kilometres in diameter, you can still appreciate its size by imagining this: 1 million earths could fit inside it. Moreover, the sun alone takes up more than 99% of our whole solar system!

- The sun is the closest thing to a perfect sphere that we can find in nature. Experts say there is only a 10-kilometer difference between the sun's polar diameter and its equatorial diameter, so considering the sun's hugeness, that difference doesn't count!

- It's not a happy scenario, but in case the sun vanished all of a sudden, we wouldn't know about it for about 8 minutes and 20 seconds. This is because the light coming from the sun takes that much time to reach the Earth. Remember: the distance of the sun from our planet is 150 million kilometres; if light travels at the speed of 300,000 kilometres per second, then we'll have 500 seconds, or 8 minutes and 20 seconds!

- Amazingly, the core of our earth is about as hot as our sun.

- Total solar eclipse, which only happens here on Earth, happens by sheer luck (or genius creation!) During this phenomenon, the sun will appear to be as big as the moon. This is because the sun's diameter is 400 times larger than that of the moon's and its distance from the earths is 400 times farther!

- As weird as it seems, according to a post by 'Fact Slides', 1/3 of Russians believe that the sun revolves around the earth.

- In photosynthesis, the plants convert the sun's rays into energy which is equivalent to 6 times the power consumption of the entire human civilization. And this happens every day, in all the plants exposed to sunlight! So just imagine the potential of solar energy. In fact, according to the inventory made by Ray Kurzweil "the world's energy needs can be met by 1/10,000th of the light coming from the sun each day".

- People who were struck with lightning experienced 5 times of the sun's surface heat!

Chapter 3: The Moon

The moon is the easiest heavenly object to find in the night sky. The time and sequence of the moon's phases has guided humankind for as long as we can remember. It has been used as a reference to measure our months and years through the lunar calendar.

Like many of its metaphors the moon has been a mystery to us even though it has been only about 300,000 km away from us. Because the moon takes about the same time to rotate on its own axis and around the earth, we are able to see only the same face of the moon, no matter where we look from.

Additionally, the moon has different faces such as the half moon, full moon and new moon. This is regulated by the amount of light it reflects off of the sun.

The moon is actually quite massive for a satellite when you compare the relative size of other satellites to other planets. It is about one-fourth the size of the earth and due to its proximity its gravitational tug on the earth is expected to have significant bearing on the sustainability of life on our planet.

Physically the moon has little or no atmosphere and this is the reason why you can even sometimes see the presence of craters on the surface even with the naked eye. Meteorites that bombard the moon rarely burn away before they reach the surface.

The origin of the moon was quite a mystery for some time and has had wide cultural historical implications. But scientists are now convinced that the moon was formed due to a huge collision between the earth and the moon millions of years ago.

They postulate that the moon had impacted the earth when it was still molten. The resulting impact caused the earth to absorb a part of the moon while the remaining tore itself away.

The theories as to origins are still debatable and there is even one saying that earth stole the moon form Venus. The moon, the sun and the earth periodically align themselves in a line in different orders that lead to the formation of solar and lunar eclipses. These eclipses have been a major area of study about the nature of light and the nature of the motion of the celestial bodies.

The moon was and is the only surface we have managed to set foot on other than the earth. What is fascinating is that although we have stopped sending humans to the moon we still have not stopped wondering about the mysteries under the surface. There is a lot of hope that we would one day be able to harvest Helium-3 in significant quantities from the moon. This could in theory solve our energy problems, as it is an extraordinary source of power.

Chapter 4: The Planet Earth

Of course, before we discuss the other planets, we need to get to know our planet: The Earth. People from many years ago were not always fascinated by Earth: for them, it was easy to accept that this is where we live and there's no more about it. Why should there be a need to learn of its characteristics, when it's already providing us with all the necessary things we could ask?

Why explore possibilities, when things are good as they were? Due to this contentment, many misconceptions arose about the earth when the time to finally explore it came, some of them are as follows:

- The earth is flat
- The earth is the centre of the universe
- The sun revolves around the earth
- The earth is not a planet

While these misconceptions are now obsolete, it is still interesting to hear about the development and cessation of the myths.

When Ptolemy created the earth-centric model, it was widely accepted. It was only the second century, so science was just blooming and people welcomed whatever new knowledge they could have.

In the model, the sun was described as the dense and unmoving centre of the galaxy – it wasn't a planet. Rather it was the culmination of everything; hence even the sun which is the ultimate source of heat and light, was believed to revolve around it. It was only when Nicolas Copernicus released his heliocentric model that the people – and the church – believed that the sun is in fact the centre of the solar system.

Still, accepting it was hard and they still couldn't face the possibility that the earth was a planet. Soon before Uranus was discovered in the 18th century, was the only time the church relented – our Earth is not unique: there are other planets like it in the universe.

But perhaps, the most interesting misconception is the flatness of the Earth. In fact, even people of today still argue (though no one knows exactly if they are just trolling the internet) that the earth is flat.

The "scandal" made a particular buzz when popular

rapper Kanye West declared in his Twitter posts, that NASA had been deceiving us all this time. The earth is flat, he said, and he even uploaded photos to prove his point. But before you think that Kanye West is the new scientist in town, keep in mind that there are others like him in his belief – they strongly disagree that the earth is round so they created the "Flat Earth Society".

If you are interested in their discussions, you can visit their website here:

http://www.theflatearthsociety.org/cms/

People from the Middle Ages knew nothing about Gravity, so it was understandable for them to think of the earth as flat – how could they possibly not fall if it wasn't? Something that's so widely believed would be hard to dispute, so just think of the scandal it caused when some people stated that the earth is spherical. The church even thought of the notion as an abomination, hence, they ostracized some people for their stand.

Pythagoras (the one with the theorem), was one of the few who thought that the earth was spherical in shape. Aside from him, Plato and Aristotle also believed the then theory. Ferdinand Magellan, a famous sailor, also fought the church on their belief, for he claimed that in his travels, he saw the shadow of the moon.

The truth is, educated people even before Pythagoras' time, knew that the earth was round! But when the news of its flatness spread like wildfire, the correct knowledge soon died.

Misconceptions aside, we now know that we have come a long, long way. We know more than the earth's shape and its position. So what is there to discover more about our home planet?

The Planet Earth at a Glance

The third planet from the sun, Earth is – as of now – the only planet with an oxygen-filled atmosphere, oceans upon oceans of water, and of course, life. With a diameter of roughly 13,000 kilometres, our planet is the 5th biggest one, larger than Mercury, Venus, and Mars, but smaller than Jupiter, Uranus, and Saturn.

Earth's shape may be round, but it's not perfect: the exact term is oblate spheroid, meaning, the polar region is slightly squashed, but the equatorial region is bulging.

This shape is brought by gravity and the earth's spinning.

1/5th of the earth's atmosphere is made up of oxygen (made by plants) which makes life possible. 70% of the surface is water. Our planet's position is not straight – it's slightly leaning; an imaginary line called axis runs from both poles. When revolving around the sun, it follows another imaginary line called orbit.

When was Earth formed? Scientists think it was around the same time our sun was born, which was some 4.5 billion years ago. Although nothing is conclusive yet, it is believed that there was no life on earth when it was created – the earliest trace of life was found in the Archean Eon (3.8 billion years ago).

You see, there were a total of 4 eons: Hadean, Archean, Proterozoic, and Phanerozoic. The first three are collectively referred to as Precambrian while the last were subdivided into three eras: The Paleozoic, Mesozoic, and Cenozoic. A lot of species appeared during the Paleozoic era, while the Mesozoic was the era of dinosaurs, and the last, the Cenozoic, was the age of mammals.

Within the Earth

One of the most fascinating things about our planet is what's beneath the surface. It has a core, which is more than half the size of the planet's diameter, about as big as Mars. The composition of the core is also interesting: the outer layer is liquid, while the inner core is solid.

Above the core, the mantle lies. Contrary to popular belief, the mantle is not composed of stiff rocks, rather, they can flow, albeit slowly. This flowing is the force that drives the movement of the uppermost layer: the crust. To have a better idea of the construction, imagine a river where pieces of wood are floating – the wood is the crust while the flowing river is the mantle. This movement ultimately causes earthquakes.

The crust can be divided into two kinds: the dry land of the continents and the ocean floor. The thickness varies depending on several factors, like the type of rocks.

As you go deeper, approaching the core, the temperature increases.

Unbelievable Facts About the Earth

- The rock on the surface (the very rock you are standing on) gets recycled after some time. The process goes like this: when there's a volcanic eruption, the volcano will spit magma, which will harden and turn into the rocks we know. In time, previously existing rocks will be sucked in by tectonic plates OR they will be pushed down by the newer rocks.

- Antarctica is a pretty amazing place, at least when it comes to ice and water. Apparently, it contains 90% of the Earth's ice AND 70% of our planet's fresh water! On top of these, this place is the coldest, windiest, and highest place on earth!

- Some parts of the Earth have less gravity than other areas. You've read that right! It means gravity is not distributed equally on earth. One place in particular is Hudson Bay in Canada. Hudson Bay has less gravity than other areas because it has less land mass.

- Did you know that the Queen of the United Kingdom owns about 1/6th of the earth's land surface?

- If there are people residing on the moon, they would tell us that they also see our earth in phases.

- The earthly soil is rich in microorganism; in fact, if you count the number of microorganism contained in a teaspoon of soil, you'll find it very populated! The numbers exceed the entire human race now!

- Just like how the dinosaurs dominated the earth before, it's interesting to note that before, there were giant mushrooms instead of trees!

- Although no other life form in space is discovered, scientists believe that in our galaxy, there are approximately 2 billion planets, like earth. So it's only possible that there are "others" out there.

- Although you seldom hear an accident involving a person being struck by lightning, you might find it scary to know that lightning strikes 100 times per second; that's more than 8 million times within a day!

- It was stated that it's only here on earth that water can be found in its three states: solid, liquid, and gas.

- Even though we round it up, the exact time it takes the earth to spin on its axis is 23 hours, 56 minutes, and 4 seconds. That means there's really no 24 hours in a day!

- The earth contains enough gold to cover its entire surface up to 1.5 feet in depth.

- A place in Texas is called Earth, and as of now, it's the only place on earth, called, well, Earth.

Chapter 5: Other Planets

The other planets which we will discuss now have varied origins and it is notable to observe that the first four planets namely Mercury, Venus, earth and Mars are all solid planets with a molten core. These planets have solid surfaces and are much smaller compared to the gas giants such as Saturn, Jupiter, Neptune and Uranus. And then there is the curious case of little Pluto, which we will discuss later.

If people from before had misconceptions about our Earth, they sure had misconceptions about other planets, too. Curious? These misconceptions are listed below:

- Ancient Greeks thought that planets were stars; they were so sure about their discovery, that they even named them “asters planetai” which means “wandering stars”.

- They (Greeks) also thought that there were two Venuses. This was due to the times of the day when they saw it – first was just before sunrise and then before sunset, hence they named them Phosphorus and Hesperus – which means "Light Bearer" and "Evening Bearer" respectively.

- Another point for the Greeks: they thought that Mars was moving backwards. To be fair, it was a good assumption since every 26 months, the earth passes Mars from behind, giving that illusion that it was moving backwards. This happens because our planet orbits the sun in about half the time that Mars does.

- Uranus was first thought as a comet. This was because when seen from our planet, Uranus doesn't shine as bright as the other stars in the sky. On top of this, scientists back then didn't notice Uranus' movement; little did they know that they had to wait longer to see the motion because the planet is larger and farther than the others.

Mercury

Mercury is characterized as the fleet footed messenger of the Roman gods and that is due to the fact that it orbits the sun in a mere 116 days. As a planet it is pretty inhospitable with temperatures ranging from 800 degrees Fahrenheit on the side exposed to the sun and about -280 degrees Fahrenheit on the dark side. That's pretty weird, right?

Just imagine our planet: it is the third planet, but it doesn't get *that* cold, so how come Mercury experiences this extreme temperatures. NASA explains that this is because of the blanket of atmosphere that we have – apparently, there's no atmosphere in Mercury, making it impossible for humans to live there. The lack of the blanket is also the reason why Mercury looks like the moon – there are a lot of impact craters.

The craters are caused by rocks falling from space; since the planet doesn't have atmosphere, the force remains great that holes are produced upon their landing. Now, why doesn't Mercury have atmosphere? This is because it has less gravity. The lack of gravity lets the atmosphere get washed into space.

Mercury is quite tricky to identify with a telescope and even more tricky to spot with the naked eye due to its proximity to the sun. Mercury is seen most simply when it is close to its greatest elongation, this is when its angular separation from the Sun is greatest.

Mercury is usually near greatest western elongation, when it is west of the Sun in the sky, so it is usually visible soon before sunrise, or utmost eastern elongation, which means it is noticeable soon after sunset.

However, the exact date of these elongations and when they are their greatest is still fairly unpredictable over the calendar year.

Research about mercury and more importantly its survival so close to the sun has been studied using research probes such as the Mariner 10, MESSENGER and Curiosity Rover. NASA has admitted on the difficulty of these missions, since no man can be sent to Mercury, but they are adamant to continue exploring using controlled ships.

Other Interesting Facts About Mercury

- If it is possible to live on Mercury, then we would be celebrating New Year every 88 days! However, one day on this planet is equivalent to 176 earth days; this is because of the gravitation lock (caused by the sun's gravity) which slows down the planet's rotation.

- Mercury has no rings, there's also no moon; again it's because of the lack of gravity.

- Due to the extremities of the temperature in Mercury, it is only the second hottest planet. Venus is the hottest even though it's farther from the sun.

- This planet is rich in iron, which causes the wrinkling of the surface. When the iron cools and dries, it will wrinkle up and can even go up to several hundred miles long.

Venus

You might have heard about Venus as the evening star (and morning star) and it is often romanticized as the goddess of beauty and it is the brightest object in the night sky with an apparent magnitude of about -4.6.

Venus has no natural satellite and its atmosphere is somewhat like a prelude to hell. It has an extremely dense atmosphere of caustic sulphur dioxide and carbon dioxide. It is reported to rain sulphuric acid and the immense amount of this gives Venus its characteristic color.

Venus orbits the sun at a typical distance of about 0.72 astronomical units and completes an orbit every 224.65 days. All planetary orbits are elliptical including our earth but Venus has a circular orbit with very little eccentricity (about 0.01). Eccentricity is a mathematical index to measure the joviality of any path.

Spotting Venus should not be a problem and there are multiple research observatories that seek to measure various physical aspects of the planet such as its rotation using methods such as Doppler shift. The Arecibo Observatory is one of such observatories.

Exploration about the chemical composition of the planet has been difficult and some of the satellites and probes haven not survived on the surface for too long because of the atmosphere.

For amateur astronomers it is next to impossible to observe the extreme landscape of Venus which is dotted with humungous mountains and deep ridges.

Other Interesting Facts About Venus

- It will take 243 earth days before Venus can complete one spin on its axis. BUT it will take only 224 days to complete one revolution! That means the years are shorter than the days!

- Moreover, it rotates in the opposite direction as most other planets. The retrograde rotation is said to be caused by an earlier collision with an astronomical body (like a comet, perhaps) that forced the direction to change. The only other planet with this kind of rotation is Saturn.

- Aside from having almost the same size as Earth, it is assumed that billions of years ago, Venus also had oceans of water on it. The problem was, it boiled up as the temperature kept on rising.

- There's too much pressure on Venus. Scientists said that the pressure on the planet is similar to the pressure one can feel if you were to go 1,000 kilometers deep into the ocean. This strong pressure is the reason why there are no craters in Venus; any falling rock or object from space is crushed before it hits the land.

- If Venus has strong pressure, the magnetic field, surprisingly, is weak. This could be due to the fact the Venus has no solid inner core, unlike our earth.

- Among all the planets, Venus has the most number of volcanoes. Scientists were able to count 1,600, but there could be more – they are just too small to notice. Most of these volcanoes, however, are dormant, but active ones are also present.

- Living on Venus is impossible due to a lot of reasons; one of which is its super windy weather. According to data, the winds are at the rate of 450 miles an hour or 724 kph! In fact, the winds are faster than the greatest tornadoes that we have here on our planet!

Mars

The planet characterized as the god of war is one of the most fascinating planets that have captured our interests in more recent times.

Mars as a planet is red in color due to it the universal deposition of iron oxide on the surface. Although it is one of the smallest planets in our Solar system and is about half the size of the earth, it has some of the most striking physical landscapes. It has both the highest mountain (Olympus Mons) and the deepest canyon or ridge (Valles Marineris) among the solid inhabitants of the solar system.

Mars is also similar to earth in many ways such as the fact that it has frozen ice caps whose size varies with its seasons, as on earth. It also has almost the same rotational period about its axis as that of the earth and one Mars day is slightly longer than ours.

One of the main reasons for the existence of life on earth is due to its optimal distance from the sun which means that the earth is neither too far nor too close.

The region that can sustain life on our solar system is known as the habitable zone and Mars lies just outside this zone; it only periodically occupies this zone during its 687 day revolution period around the sun.

Couple this with its thin atmosphere and temperature fluctuations; it is almost impossible for intelligent life or even life to grow on Mars. That being said there are some examples that suggest that life might have existed on the planet. These facts have not stopped the predominance of "Alien life forms from Mars" is an all too popular theme, in our culture.

It has two satellites namely: Phobos and Deimos, which are actually two captured meteorites orbiting the planet.

Mars usually appears to the naked eye as distinctly yellow, orange, or red; the actual color of Mars is closer to that of the color of butterscotch ice-cream. It has an elliptic orbit so its magnitude of luminosity varies. One interesting fact is that Mars exhibits retrograde motion which means that it can appear to move backwards in looping motion as you approach closer it.

There are about five operating spacecraft revolving around Mars and two operating on the surface with more scheduled to join them.

Other Interesting Facts About Mars

- Like earth, Mars has water, too, albeit, frozen – but it doesn't mean that you can live there because other factors also count, like atmosphere (it's composed of 95% carbon dioxide!), landscape, weather (temperature is really, really cold there), etc. As mentioned, this planet is still outside the habitable zone. The planet is extremely dangerous for an unprotected spaceman.

- One of the main factors why Mars has the tallest of the mountains is because it has less gravity. Even though the mountains are tall, there's less risk of them collapsing.

The Gas Giants

A gas giant is any planet that is not made up of mostly rock and other solid substances. The term was created by James Blish, a science fiction writer from the mid-1900. These planets are primarily made up of various gases ranging from hydrogen to chlorine.

Physically though it is not possible to have a gigantic ball of gas to be a planet without something dense and solid or semi-solid at its centre to form it gravitational nucleus, so is the reason that the gas giants have a core made of liquid compounds and molten rock.

Gas giants are also called Jovian planets after Jupiter, which is considered as the model of gas giants in our Solar System. There are altogether four gas giants present in the Solar System. They are Jupiter, Uranus, Saturn and Neptune.

The gas giants in our Solar Systems share a number of parallel characteristics. All the gas giants and planets after Mars are considered as outer planets. They are also significantly larger than all terrestrial planets put together. Jupiter for example is about 318 times larger than earth. Gas giants are also hoarders of the moon. These moons are captured comets that pass too close to these beasts.

Gas giants are also extremely light and density is very low and it is theorized that Saturn could probably float on water! The satellites of Saturn are very interesting as some of them are thought to have the potential to harbor life underneath their frozen surface of water and ice.

To observer gas giants through a telescope is one of the most enthralling experiences for an astronomer. This is because they have the most striking features imaginable. Below are more facts about the Gas Giants in our Solar System.

Jupiter is known for the Giant Red Spot which is thought to be a raging storm about thrice the size of the earth. If our earth has one moon, then Jupiter has 50 and NASA said that there are still 17 waiting to be confirmed! This huge planet is mostly made up of hydrogen and helium (which are poison) and has an icy surface on top of rocky structures.

Should you decide to visit this planet, you are in for a tough experience, and honestly, you might die even with protection because you would melt. The pressure on Jupiter is also no joke – in fact, it's too much that it can turn gas into liquid! Unknown to a lot of people, Jupiter also has three rings, but they are not that visible and not nearly as beautiful as what Saturn has.

Saturn is of course famous for its glorious rings, whose origins are thought to be the remnants of an earlier planet. This planet was discovered by William Herschel in 1781 and is about 9 times larger than our home planet's radius. Like Jupiter, the atmosphere is made mostly of poison: hydrogen and helium, making human life impossible to sustain.

Saturn has more moons than Jupiter (53) and 9 more on the waiting list, although if the other 17 moons of Jupiter are confirmed then it will have 67 moons. Although Saturn cannot support life, NASA said that some of its moon probably could.

Uranus and Neptune are collectively known as ice giants and are noted for their striking blue hues. All the gas giants up to Neptune can be appreciated through good quality telescopes.

Uranus is also called the sideways planet due to its

almost parallel tilt. According to experts, a collision from an earth-sized object might have caused this unique tilt. Helium, methane, and hydrogen are the main composition of Uranus' atmosphere, with methane causing the blue hue. Although one day in Uranus is only 17 earth hours, the years are long: one complete revolution will take up to 84 earth years!

Interestingly, all 27 moons that Uranus has are named after the works of William Shakespeare and Alexander Pope. Lastly, this planet has 13 rings – with the inner rings narrow and dark in color, and the outer rings featuring bright hues.

Neptune is another interesting planet which just completed its first revolution in 2011 since it was first discovered in 1846. One day on this planet is just 16 hours, but one year is equivalent to 165 earth years! Aside from helium and hydrogen, it also has a little amount of methane which gives its blue color; that's why Neptune and Uranus are known as sister planets.

This planet has 6 rings, and 13 moons all named from sea gods in Greek Mythology. Although it doesn't often happen, sometimes, Pluto "shares" Neptune's orbit, making the dwarf planet closer to us and the sun than

Neptune. This is also one of the reasons why NASA has demoted Pluto from being a major planet, to a dwarf. More about this demotion will be discussed later.

Gas giants are also a little eccentric like Jupiter which rotates so fast that it is actually flattened at the poles. Neptune and Uranus are practically twins and Uranus is actually tilted so much about its axis that it appears to roll as it travels around the Sun!

Gas giants are among the most prevalent types of planets in the entire universe and almost all our information comes from the still functioning Voyager spacecraft which is still soldiering farther than it was ever intended to.

Pluto

Pluto has always been an oddity in our solar system and was excommunicated from the planet family and is now classified as a dwarf planet and is the newest planet to be discovered. It is the only planet to not be named after a mythical god or goddess and is also the coldest known place in our solar system.

It's interesting to discuss how Pluto was "dumped" by NASA as a planet, so, we will cover the characteristics that a planet should have:

- A planet should be able to revolve around the sun
- A planet should have enough gravity to pull itself together as a spherical mass
- A planet should have its own orbit

If you'll notice, nothing is wrong with Pluto! It revolves around the sun so it must have its own orbit AND it's spherical in shape, meaning it has enough gravity. The problem is, NASA found out that Pluto doesn't really have its own orbit, or more specifically, it doesn't have a clear neighbourhood around it.

How come?

According to them, when a planet is formed, it becomes the strongest gravitational pull in its area (orbit). If an object comes near it, it either becomes sucked into the planet, or ejected out. Further investigation revealed Pluto wasn't able to do that. Apparently, there are other objects around it that wasn't ejected or sucked in, hence, it became a dwarf planet.

Comets, Asteroids, Meteors, and Meteorites

There is a belt of asteroids that lie between Mars and Jupiter and is a relatively unexplored belt in our solar system. There are special classes of asteroids known as Trojans which orbit planets without colliding with them because they lie in points known as Lagrangian points.

Jupiter is known to have the most "Trojans". Meteorites are the remnants of asteroids after they pass through the atmosphere and they are among our biggest sources of information about the composition of rocks outside of earth. Asteroids are thought to have large deposits of rare minerals and asteroid mining might be a future entrepreneurship exploit.

Comets are the enigmatic nomads that orbit the sun. They are beautiful in their appearance but it is ironic to know that much of their splendour is due to their slow death. Some of the famous comets include Comet Hale-Bopp and Halley's Comet which are noted for the large and bright burning nucleus. NASA has landed on comets through the Rosetta mission and it seeks to understand more about the solar system as its passenger.

Despite these majestic descriptions, what is really the

difference between comets and asteroids? What are meteors and why do we still have meteoroids and meteorites?

Asteroids – are just chunks of rocks which came from the Asteroid Belt we have mentioned. Sometimes, though, they come close to the sun, and as a result, they can also get near our planet.

Comets – are somewhat like the asteroids, but they have more "ingredients", like ice, methane, and ammonia, and other compounds. All these give the structure a "fuzzy" appearance – called coma – and a tail, once it gets near the sun.

With these two cleared up, the next in the discussion is the difference between meteoroid, meteors, and meteorites:

Meteoroid – is like an asteroid but way, way smaller. It's just a chunk of rock that hasn't yet entered our atmosphere.

Meteor – is a meteoroid that falls into our earth. The pressure from the atmosphere as it falls gives us the appearance that it's a "falling star". Most of the time, the meteor disintegrates due to the atmospheric pressure.

Meteorite – is a meteor that landed successfully here on earth. Because of the pressure, the size will be considerably smaller.

Chapter 6: The Other Entities

Astronomy is an enormously huge field with libraries dedicated to its exploration. Astronomy is extensively studied through other modes of observation such as radio and infra-red telescopes and majority of the stunning images we have are courtesy of the Hubble and the X-Ray enabled Chandra observatory.

Black Holes

Black holes are a consequence of Einstein's theory of relativity. These are massive stars that have collapsed within themselves beyond a critical radius such that they become so dense that it starts to exert enough force to pull light into its mysterious inner workings.

The point near a black hole beyond which there is no escape is called the event horizon. Black holes are also thought to be at the centre of our galaxy. There are three types of black holes: the stellar, the supermassive, and the intermediate.

The Stellar Black Holes

Stellar black holes are those stars that continue to collapse unto itself. You see, for an ordinary, smaller star, using all its fuel means it would collapse until a core is formed, which will then be the white dwarf or the neutron star. However, if the star is big, then it won't stop collapsing, creating a stellar black hole.

Although these black holes are small (just equivalent to three times the mass of our sun), they are not less deadly. The resulting gravitational force can "suck" anything that comes around it. Frighteningly, NASA said there are about "a few hundred million" stellar black holes in our galaxy alone.

The Supermassive Black Holes

Although there are a lot of Stellar Black Holes, the ones that dominate our universe are the supermassive kind. They are supermassive because their mass could be millions or billions of times larger than our sun! It's not clear how these types of black holes are formed, but scientists believe they are a result of many black holes merging together.

The Intermediate Black Holes

If the black hole is neither small nor supermassive, then chances are it's of the intermediate kind. At first, scientists believed that black holes only came in small or super huge variety, but now they know that they were wrong.

How Do Black Holes Work?

As you have already learned, NOTHING can escape from these deadly entities, not even light. Due to this, observing black holes is difficult for scientists – they can't see them, so they have to rely on the radiation it emits as the dust and other particles get pulled in it.

The event horizon can be divided into two more layers: the outer and the inner ones. One more layer is present, which is called singularity. If the event horizon is the area where nothing can escape, the singularity is where the mass of the black hole is concentrated.

Is it really impossible for anything to escape a black hole? Ordinary physics has it that, yes – nothing can. However, if quantum physics is taken into consideration, things can drastically change. Quantum mechanics, as we all know, deals with atoms, and according to it, every particle has an anti-particle: something with similar mass, but with negative electric charge.

These two should not meet, because if they do, they can destroy (annihilate) each other. If a particle-anti-particle pair forms in an area just outside the event horizon, one of the particles can get sucked in, while the other can be ejected.

This will cause the event horizon to shrink, which is impossible, according to classical physics. To date, scientists are still struggling to completely understand how black holes work exactly.

Other Interesting Facts About Black Holes

- In case you are unlucky enough to get sucked into a black hole, scientists say your death will be swift – you are dead even before you reach the area of singularity. Either the gravity will stretch you out like spaghetti or the event horizon will burn you up.

- Despite the fact that we used the term "sucked" here a couple of times, black holes don't really do that. Instead of pulling them in, objects fall into them.

- Black holes are so powerful that if a star passes by it, the star can be torn apart.

Wormholes

One consequence of the black hole is the wormhole which thought to be a connection between two black holes and the properties of these wormholes allow you to time travel. This is due to the fact that the bend in the space time fabric in a way, allows for a short cut.

To understand this, you need to imagine the three dimensions of space and one dimensional time to be intertwined into a fabric and if you bend this fabric, like say a pizza in the shape of a U, the wormhole is the shortest distance between the two ends and the normal path we would take would be along the U curvature of the pizza.

The curvature is a natural result of gravity. The more mass an object contains, the more it will tend to bend space-time and this effect is compounded if the object is small.

Hence the almost immeasurable magnitude of the force of gravity of black holes bend space time in on itself to and it theorized that sometimes one black hole can do this to the extent of connecting with other black holes on the other ends of the universe.

What does it feel like to be inside a wormhole?

You might have read a book or watched a movie that features wormhole travelling. Don't believe them yet – first of all, science has yet to find a wormhole to really investigate what it would be like to be in it.

Second, it's impossible for anyone to travel a wormhole since, according to science, they are relatively small. As in microscopic small (10 to the -33rd power centimeters). Still, enthusiasts argue that since our universe is still expanding, it's possible that wormholes, too, have grown.

The next problem one may experience in case he wants to travel through a wormhole is its stability. Apparently, wormholes can collapse unless they contain enough "exotic" matter. If it does, then it will remain open and unchanging for a longer period of time.

But what exactly is an "exotic" matter? It's not a dark matter, nor is it an antimatter – it's something that has a lot of negative pressure and negative energy. Is it naturally occurring? Perhaps. Can we make it? Maybe.

If ever in the future we find out that wormholes are real, and that they could be travelled, then scientists assume that they can connect two universes.

Dark Matter

The concept of dark matter is very intriguing and is actually an interdisciplinary consequence of astrophysics and astronomy. For astronomers it is very hard to observe dark matter even though it is thought to make up the majority of the universe.

It neither absorbs nor emits electromagnetic radiation and it can only be measured by looking for its apparent effects. The presence of the invisible mass of dark matter is needed for understanding of the space-time curvature of the universe.

Dark Energy, simply put, is the energy that's causing the rapid expansion of the universe. You see, before, scientists thought that the expansion was slowing down due to gravity – how could our universe possibly continue to expand if a lot of objects in space have a gravity to stop the movement?

However, in 1998, while observing a supernova, scientists gather that billions of years ago, our universe was expanding more slowly than it is now.

We don't know much about this mysterious dark matter, but one thing is for certain, scientists won't stop digging.

Antimatter

We have mentioned about the anti-particle earlier in the discussion of black holes, but we didn't mention that if anti-particles joined together, an antimatter will form. As such, they have the same characteristics: same mass, but with different electric charge.

In theory, both matter and antimatter were formed after the Big Bang happened. But wait; if that was the case, then life couldn't have possibly existed, right? After all, when anti-matter and matter meet, they would annihilate each other, leaving nothing but energy behind. So what happened?

Life as we know it became possible due to a rare asymmetry. Apparently, for every one billion matter-antimatter pair, there is an extra matter particle, which collided with nothing, hence, it survived, making life possible.

Science has also determined that antimatter is closer to us than we've ever imagined. In fact, they can even be found in bananas! The only thing saving us from being annihilated is the fact that the small antimatter particles are short-lived, meaning they pose no dangers. In fact, they are even used in medicines!

Throughout the years, scientists and physicists were able to study antimatter because of Penning Traps or antimatter trap. In this device, the antimatter floats because of a magnetic field, preventing it from colliding with the walls, which are matter.

Chapter 7: The Galaxies

Galaxies are a collection of dust, particles, and stars, that's why others call it "Star Cities". It's virtually impossible to count all the galaxies that we have, since in our universe alone, we might have one hundred billion; some like our Milky Way, while others are more different.

Parts of Galaxies

Our galaxies have 5 parts (depending on kind): the bulge, the disc, arms, halo, and stars, dust, and gas.

The Bulge – is often at the center of the galaxy; as what the name suggests, it's round in shape and bulging on top. The bulge is wide – it can reach up to 10,000 lightyears across.

The Disk – is found on spiral galaxies. This is the pancake-shaped region surrounding the bulge. The disk is also huge: it's 100 lightyears across and about 1,000 lightyears thick.

The Spiral Arms – are only found in spiral galaxies. The

arms emanate from the bulge itself, then rounds up or down to give the galaxy a pinwheel appearance. Most of the stars here are blue stars, or the young ones.

The Halo – can be found just outside the bulge; in fact, it's kind of hard to distinguish between the outer layer of the bulge and the halo. This area contains mostly clusters of old stars. Halos typically have dark matter, and can have a measurement of 130,000 lightyears across.

Stars, Gas, and Dust – of course, you cannot call a galaxy a star city if it doesn't have stars. Dust, gas, and planets come along with it.

There are three major types of galaxies: spiral, elliptical, and irregular; let's discuss them one by one.

Spiral Galaxies

The Milky Way is an example of a spiral galaxy; they usually have bulging center and spiral arms. The planets, dust, particles, and stars inside the galaxies revolve around the galactic center at a regular pace: hundreds of kilometres per second, though we might not feel it since we are mere dots in space. Most of the new stars can be found on the spiral arms, while the old ones are situated at the galactic center.

Elliptical Galaxies

As the name suggests, elliptical galaxies are elongated in shape – like a cigar. Unlike the Spiral Galaxies, the elliptical kinds often have old stars – up to one trillion – and fewer new stars. This kind of galaxy is also often smaller in size than spiral and irregular.

Irregular Galaxies

Irregular galaxies do not fall in the spherical shape and elliptical shape.

Milky Way, Our Galaxy

We cannot discuss the galaxies without mentioning the Milky Way, our galactic home. It has 200 billion stars; our sun is just one of them and it's not even one of the biggest. If you'll ask a scientist, he'll tell you that our galaxy is just a mid-weight city – other galaxies have up to 100 trillion stars! This is the reason why there's so much to explore, and there is certainly hope for another life form.

As mentioned, the shape of the Milky Way is spiral, it has a halo, a bulge, arms, and a lot of gas and dust. Unknown to a lot of people, our galaxy is a product of "melting" other smaller galaxies which crossed our path as it moves.

Although we won't be able to experience it, in about 4 billion years, the Milky Way is bound to collide into the nearest major galaxy: The Andromeda.

Other Interesting Facts About the Milky Way

- It's still impossible to take a picture of our galaxy from above; the ones you see in books are just the artist's interpretation, or they have used a model galaxy which isn't ours. As of yet, we still cannot take a picture of Milky Way from above due to the distance: we are located at the disc, which is about 26,000 lightyears away from the surface of the galaxy. It's like taking a photo of the top of your house when you can't even get passed its roof.

- At the very heart of our Galaxy is Sagittarius A, which is believed to be a big black hole.

- Funnily enough, astronomers from the past thought that ALL the stars in the universe are all in the Milky Way. It might have come as a surprise to them when they discovered how little our galaxy is compared to the universe.

- The Methuselah Star is the oldest in our Milky Way home. In fact, it's alive for as long as the Milky Way has been in existence.

Andromeda, Our Neighbor

While it is true that there are dozens of neighboring galaxies, Andromeda is the only major one that's closest to the earth. Did you know that it's the only galaxy that we can see here on earth?

The Andromeda – also called the M31 – can be noticed from the mid-northern latitudes every night, all year long, although a lot of people only see them in northern autumn because that's when M31 is high enough for our skies.

Ancient astronomers, as usual, made mistakes in studying this galaxy: they thought of it as a collection of dust and glowing gases, hence they named it The Great Andromeda Nebula. If not that, then they thought it was another solar system in the making. After they recognized it as a spiral nebula, they argued if it was situated within our galaxy or outside.

In the 1920s, the Andromeda was finally recognized as a galaxy that's far away yet still closest to our home. In the Local Group of Galaxies (which has 54 galaxies, most of which are minor ones), the Milky Way and the Andromeda are the two massive dominants.

Other Interesting Facts About the Andromeda Galaxy

- It's named after the area where we can see it from here on earth, the Adromeda constellation, which in turn, is named after a Greek mythological princess

- Just because the Milky Way contains more dark matter doesn't mean that it's the most massive – the truth is Andromeda appears to be the most massive among the galaxies in the Local Group. In 2006, scientists discovered that the Milky Way is just 80% of the Andromeda's total mass.

- On top of that, the number of stars in contains can reach up to 1 trillion, compared to the Milky Way's 200 billion stars (although some speculate that there could be as many as 400 billion)

Chapter 8: The Infamous Apollo 11 Mission

On July 20, 1969, mankind made history: they were able to land on the moon successfully. People cheered and marvelled at the data the crew took home, but it doesn't mean that the infamous Apollo 11 Mission escaped controversies.

It had 3 main crews: the commander, Neil Armstrong, Michael Collins, the command module pilot, and Edwin Aldrin, the lunar module pilot. There were also three backup crew, 9 support crews, and 3 flight directors.

The launch of the mission itself was such a highlight: people from the area of the launching crowded the place and millions of people watched it happen from their television, including then president Richard Nixon, who watched it from his Oval Office at the White House.

The setting of the launch was at the Kennedy Space Center; it took place on July 16, 1969. The rocket that launched the mission was Saturn V (pronounced as Saturn Five) from Launch Complex 39 site, particularly, the Launch Pad 39. 12 minutes after the ship was launched, it reached the Earth's orbit. After the successful flight to space, the next milestone was its lunar descent and the infamous moon landing.

At first, everyone was ecstatic with the success of the mission, but after a little while controversies surfaced:

Why was the flag of the United States flapping, as if wind is present, when in fact, there shouldn't be air?

Space Historian Roger Launius said that the wind didn't cause the flapping – it was the inertia force from when the astronaut placed it there.

Weren't there only 2 astronauts at a time, so who took the photos where both astronauts were visible?

The photo in question features an astronaut's head; on his helmet, the reflection of another astronaut was visible. How was that possible? Who was holding the camera?

While the argument is valid, it can be explained. The presence of two astronauts doesn't mean that there was a third person in the area, nor does it mean that the landing was a hoax. Apparently, the camera was attached to the astronauts' chest.

Didn't they leave mementos in space? Including the seismometer? Those objects should still be there, right? So why don't future photos feature them?

The objects could still be there, however, no camera has a resolution that will capture the mementos, unless of course, another mission is sent to moon.

The footprints were so clear? Shouldn't they be fuzzy because the sand is bone-dry?

This question was asked in particular because of the photo which features Aldrin's footprints, which are so clear. Space experts said that the claim that the prints were stepped on wet sand is pointless – the sand in space is dry, but it's also powdery. And since there's no air on the moon, the pristine-like print will stay there for a long time, undisturbed.

Are those stage lights?

When photos were taken, it was as if the surrounding was

lighted, so people assumed that the landing was taped in a studio. Experts said it was impossible! Firstly, if NASA spent millions of dollars to stage the event, then they wouldn't have made such an elementary mistake! According to them, those irregular lighting patterns were simply caused by lens flares.

After this explanation, the controversy-believers still insisted that there shouldn't be light coming from multiple sources. In one photo, you can see Aldrin standing over the shadow, yet he is still clearly visible, which means that there is still light somewhere else.

But the defenders argue that there were multiple light sources: you have the sun, the light reflected from the Earth, the light from the lunar module, and from the spacesuit. They also explained that the surface on the moon is not flat; hence, if you are in a dipped surface, then the shadows would truly be irregular.

If we can see stars from here on earth, why couldn't you?

One of the most interesting (and convincing) points of the argument was this: all the photos released by NASA featured black skies – as if there were no stars in space.

They said that, if from the earth we can see stars with our naked eye, how come their camera didn't capture at least one spark?

NASA explained that on the moon, there were glares caused by the sun's reflection on the surface; this glare makes it difficult for them to see the stars. Secondly, the cameras they used were on a fast setting, which took photos at $1/150^{th}$ or $1/250^{th}$ of a second, making it impossible for them to capture the stars.

There was no dust...

One particular photo sparked the interest of skeptics – it featured the lunar module, Eagle, sitting prettily on the moon's surface which was undisturbed. The same skeptics said: if the module truly landed, then there should be dust floating around it from the impact.

NASA said that science fiction is to be blamed for such an argument. In comics and other movies, you can see a spaceship landing with force, but that wasn't the case when they landed on the moon. The module was controlled or throttled before landing, significantly decreasing the impact.

Shouldn't the astronauts be fried?

Skeptics said that above us, there's the Van Allen Radiation Belt which should have fried the astronauts to bits because of the increased amount of radiation. NASA, however, said that the module only stayed in the vicinity of the belt for a short amount of time, making the amount of radiation insignificant enough to cause danger.

From the questions and answers above, you can say that NASA has been defending their side pretty well; the problem is this: some skeptics still refuse to believe the landing and even found other argument points which NASA seemed to be having difficulty to counter back.

For example: if the landing wasn't really staged, then why did it seem like they just "mimicked" the low-gravity condition on the moon? For them, it was apparent in how the astronauts walked.

Apparently, if you adjust the setting of the footage and put the speed at x2.5, the astronauts would appear to be walking here on earth. As for the heightened jumps which are impossible to perform here on earth, skeptics said there were hidden cable wires. In fact, they even saw the wire's outline, even though they admitted to them being vague.

NASA has yet to dispute this claim, but to be fair, it's kind of impossible to dispute a claim as vague as this. If the skeptics chose to see unclear and hidden cable wires, then what could they do? Adjusting the speed of the footage is also out of their hands.

Another popular argument was the "C" Rock, so aptly named because there's a perfect letter C on its right side which seemed to be engraved. The thing is, the perfection of the letter made it impossible to accept that it was a natural occurrence, suggesting that somewhere along the production process, it had been marked by one of the crew.

This time, NASA had difficulty explaining the event. Their excuses also seemed suspicious: first, they claimed that the photo developer must have made this as a form practical joke, then later on they suggested that it was just a stray hair which got tangled during the photo's developing process.

Could NASA truly have hoaxed the momentous event? There will always be two sides in this story, but perhaps that is just the way it has to be to make things more interesting.

Conclusion

Thank you again for purchasing this book!

I could not possibly have hoped to explore all the fields of astronomy within in the space of this book. Astronomy is as much a theoretical study as it is a practical exercise in patient observation.

There are multiple inter disciplinary sciences involved such as astrobiology, astrophysics and astrostatistics. All these topics and each individual sub section mentioned in the book can be taken as a separate area of specialization. Also there are fields such as inter galactic astronomy that might interest the reader further!

The evolution of astronomy and man's intellectual development has been extremely intertwined. The discoveries of astronomy and its associated studies have had their repercussions in various fields ranging from engineering and physics to religion and philosophy.

To start your journey into the universe I recommend you to buy a telescope and binoculars and observe the night sky on a regular basis. Note down patterns and movements of the stars, constellations and planets over time and gain an appreciation over the difficulties our ancestors must have faced.

It is also advisable to read upon popular theories such as The Laws of Motion by Newton and The Theory of Relativity by Einstein to know and understand how the cosmic entities behave the way they do.

NASA is among the foremost astronomical establishments in the world and heavily relies on amateurs to make new and interesting discoveries. It also has a very vibrant and informative news section that continually updates the public about eclipses, meteor showers, spacecraft launches and recent photographs taken by its numerous observatories.

True justice can only be done if you start looking into the sky on your own and discover the wonders for yourself!

Thank you again for purchasing this book! If you enjoyed this book, would you be so kind as to leave me a review on Amazon? Thank you so much, it is very much appreciated!

Do You Want More Books?

How would you like books arriving in your inbox each week?

They're FREE!

We publish books on all sorts of non-fiction niches and send them to our subscribers each week to spread the love.

All you have to do is sign up and you're good to go!

Just go to the link below, sign up, sit back and wait for your book downloads to arrive.

We couldn't have made it any easier. Enjoy!

www.LibraryBugs.com

Made in the USA
Middletown, DE
16 September 2018